SAVING EARTH'S BIOMES

SAVING THE OCEANS FROM PLASTIC

by Rachel Hamby

FOCUS
READERS.

NAVIGATOR

WWW.FOCUSREADERS.COM

Focus Readers is distributed by North Star Editions:
sales@northstareditions.com | 888-417-0195

Produced for Focus Readers by Red Line Editorial.

Content Consultant: Susanne M. Brander, Assistant Professor of Environmental and Molecular Toxicology, Oregon State University

Photographs ©: Magnus Larsson/iStockphoto, cover, 1; Lotus_studio/Shutterstock Images, 4–5; Rich Carey/Shutterstock Images, 7; Red Line Editorial, 9; D.Romeo/VWPics/Newscom, 10–11; Loretta Sze/Shutterstock Images, 13; Gautam Singh/AP Images, 15; Michael Patrick O'Neill/Oceans-Image/Photoshot/Newscom, 17; Wayhome Studio/Shutterstock Images, 18–19; DragonImages/iStockphoto, 21; Belen Strehl/Shutterstock Images, 23; FotografiaBasica/iStockphoto, 24–25; smereka/Shutterstock Images, 27; petovarga/iStockphoto, 29 (labeled arrows, recycling bin, and factory), 29 (bottles).

Library of Congress Cataloging-in-Publication Data
Names: Hamby, Rachel, 1970- author.
Title: Saving the oceans from plastic / by Rachel Hamby.
Description: Lake Elmo, MN : Focus Readers, [2020] | Series: Saving earth's biomes | Includes index. | Audience: Grades 4-6
Identifiers: LCCN 2019031750 (print) | LCCN 2019031751 (ebook) | ISBN 9781644930717 (hardcover) | ISBN 9781644931509 (paperback) | ISBN 9781644933084 (pdf) | ISBN 9781644932292 (hosted ebook)
Subjects: LCSH: Plastic marine debris--Environmental aspects--Juvenile literature. | Plastic scrap--Environmental aspects--Juvenile literature. | Waste disposal in the ocean--Juvenile literature.
Classification: LCC GC1090 .H36 2020 (print) | LCC GC1090 (ebook) | DDC 577.7/27--dc23
LC record available at https://lccn.loc.gov/2019031750
LC ebook record available at https://lccn.loc.gov/2019031751

Printed in the United States of America
Mankato, MN
012020

ABOUT THE AUTHOR

Rachel Hamby writes poetry, short fiction, and nonfiction for kids. She was born on the very first Earth Day in 1970. She enjoys writing books about the environment and hopes to inspire young scientists to ask questions and find answers.

TABLE OF CONTENTS

CHAPTER 1

PLASTIC IN THE OCEANS

Earth's oceans are home to a wide variety of **species**. Crabs and sea stars crawl on the ocean floor. Fish and whales swim in the open waters. **Algae** and plants grow in kelp forests.

The ocean is one massive body of salt water. Earth's continents split this body of water into five smaller oceans.

Scientists believe there are at least two million animal species in the oceans.

The marine **biome** includes all five oceans, as well as many smaller gulfs and bays. This water is the largest biome on Earth. It covers 71 percent of the planet.

Oceans have many ecosystems. These systems are communities of living things that interact with their surrounding environment. Some of the environments are shallow and sunny. Deep waters are mostly dark and cold. But there is life in every ocean environment.

People depend on the oceans for food and other resources. However, humans have negatively impacted the oceans. One massive problem is plastic pollution. Every year, more than 8 million tons

Most ocean debris is made of plastic.

(7.3 million metric tons) of plastic enters the oceans. That amount is equal to a garbage truck full of plastic every minute.

Plastic enters the oceans in many ways. For example, plastic litter is waste that people drop or leave behind. Wind and rain can carry the waste into drainage systems. These systems wash into rivers. The rivers flow into oceans.

WHAT IS PLASTIC?

Most plastic is made from natural gas or oil. To produce plastic, oil is separated into different chemical compounds. Then, some of these compounds are made into **polymers**. Polymers are liquid when hot. They are solid when cooled. Plastic items are made with polymers, sometimes mixed with other chemicals. These items take hundreds of years to decay or decompose. Nearly all the plastic ever made still exists today.

Once in the oceans, plastic moves with ocean currents. Large amounts of waste collect in areas where winds and currents swirl together. These masses of waste are called garbage patches. Ocean garbage patches spread out over many miles.

Plastic in the oceans can kill animals or make them sick. It affects the growth of plants. It may be harming people, too. Saving the oceans from plastic is important for a healthy planet.

OCEAN CURRENTS AND GARBAGE PATCHES

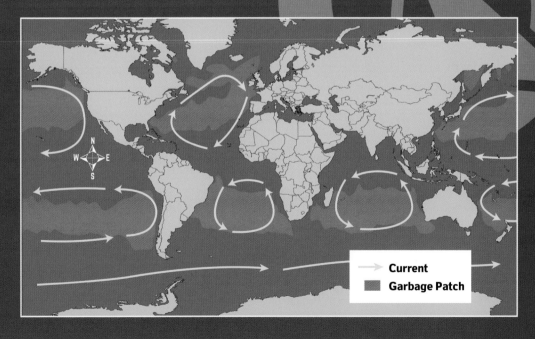

Current

Garbage Patch

CHAPTER 2

PLASTIC POLLUTION

Plastic pollution can harm ocean animals in various ways. For example, fishing teams sometimes lose their plastic nets in the ocean. These nets can trap ocean **mammals** such as whales and dolphins.

Other animals eat small pieces of plastic, thinking the pieces are food.

A rescuer tries to free the tail of a whale that was trapped by plastic nets.

The plastic can get stuck in the animals' stomachs. Then, the animals feel full. As a result, they may starve because they do not eat enough food.

When ocean animals eat plastic, humans are affected, too. A lot of ocean plastics are in the form of microplastics.

MICROPLASTICS

Microplastics are less than 0.2 inches (5 mm) long. Some microplastics are made to be this size. They are part of many household items, such as glitter and paint. Other microplastics come from larger plastic items. These items include bags, tires, and cups. In the ocean, sunlight and waves break these items down. Over time, they break up into smaller and smaller pieces.

Microplastics have been found nearly everywhere on Earth.

Some types of **plankton** eat these plastics. And larger ocean life relies on plankton for food. Small fish eat it. Other ocean animals eat those fish. And people around the world eat seafood.

Microplastics may remain in the animals that eat it. As a result, many people eat plastic. The chemicals added to plastic may be harmful to ocean life and people. While in the ocean, plastic can attract more chemicals to it. These chemicals may also be harmful for humans.

In addition, plastic pollution threatens ocean habitats. For example, plastic is harming many mangrove forests. The trees of these ocean habitats grow in warm areas where water meets land. The forests are home to many types of animals. However, when plastic gets caught in mangrove tree roots, the trees

Plastic covers a tree in a mangrove forest.

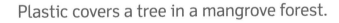

have trouble breathing. As a result,
mangroves become less able to support a
wide variety of life.

THE LEATHERBACK SEA TURTLE

Seven species of sea turtles live in the oceans. Most turtles have hard, bony shells. But leatherback sea turtles have soft, rubbery shells. These turtles are also the largest on Earth. Adults can grow to be 6 feet (1.8 m) long. And they can weigh up to 2,000 pounds (900 kg).

Female leatherbacks lay eggs every one to three years. They lay more than 500 eggs each nesting season. But fewer than 10 of these eggs usually grow into adults. Other animals eat most of the turtles' eggs and babies.

Plastic pollution is a danger to young turtles of all species. However, adult leatherbacks are also at risk. Their diet includes jellyfish. The turtles often confuse plastic waste, such as balloons,

Leatherback sea turtles live in a variety of climates, from tropical to arctic.

bags, and plastic film, with jellyfish. Eating this plastic has killed many leatherbacks. Plastic pollution may be one reason why leatherback turtles are **endangered**.

CHAPTER 3

REDUCING PLASTIC

When scientists found plastic in the oceans, they discovered the waste was causing a great deal of harm. Their findings made people aware of the problem. As a result, people started looking for solutions. For example, many groups started picking up plastic waste on beaches.

In 2017, people picked up more than 1.5 million plastic bottles from beaches.

In addition, people started recycling some types of plastic. However, people have to remember to do it. The process also costs a lot. New plastic is often cheaper for companies to buy. And some countries cannot pay for recycling programs. For these reasons, less than 10 percent of all plastic gets recycled.

Plastic can also be reused. For example, people can reuse plastic shopping bags as garbage bags. They can use plastic containers again, too. Some plastic items are made to be used more than once.

Most importantly, people can reduce their use of plastic. For instance, liquid

Eating ice cream in a cone instead of a cup is one way to cut down on plastic.

soap comes in plastic bottles. The empty bottles can end up in the oceans. Instead of this type of soap, people can buy bars. They can also refill liquid soap containers in some grocery stores. Using less plastic is by far the best way to stop ocean plastic pollution.

Most people have not cut down on their use of plastic. In fact, plastic production has increased over the years. Since 1980, it has grown more than 500 percent. And scientists predict that production will grow much more by 2050.

Changing individual people's plastic habits is not enough. Larger changes are needed. Some cities and states have banned single-use plastic items, such as plastic straws. These plastics are made to be thrown out after only one use.

In 2002, Ireland started taxing plastic bags at stores. This tax made it cheaper for people to bring their own reusable bags. Within one year, Irish people's use

Between 2002 and 2012, Irish people's plastic bag use dropped from an average of 350 bags to 14 bags per year.

of plastic bags dropped by more than 90 percent. Since then, many other places have taxed plastic bags. This type of change helps cut plastic use on a large scale. Still, much more needs to be done to stop plastic from polluting the oceans.

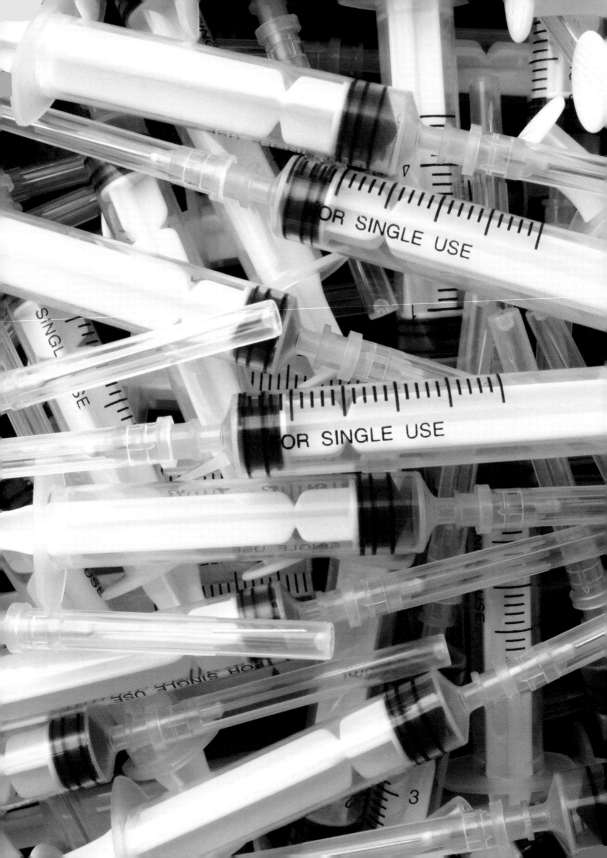

CHAPTER 4

LOOKING AHEAD

Plastic has become an important part of human life. From computer chips to cars, many useful things are made with plastic. Some products, such as syringes, help save lives. But plastic in the oceans harms the planet. More people know about the problem now. And more changes are taking place.

Single-use syringes help prevent the spread of infections.

Individuals continue to use less plastic. They continue to reuse and recycle plastic. But they are also starting to refuse it. For example, some people refuse plastic forks, knives, and straws at restaurants. More governments are also banning single-use plastics. Several states have banned plastic bags in

MR. TRASH WHEEL

The city of Baltimore, Maryland, lies next to a harbor. This harbor flows into the Atlantic Ocean. But the harbor is polluted. In 2014, the city put a new type of machine in the water. This machine collected waste. It was called Mr. Trash Wheel. By 2019, the machine had collected more than one million plastic items.

Mr. Trash Wheel floats in Baltimore Harbor.

most grocery stores. As of 2018, more than 60 countries have taxed or banned single-use plastics in stores.

Companies are also making changes. Some companies have stopped using plastic packaging. Nearly half of all plastic waste comes from packaging. Bulk food options at stores are increasing, too.

With bulk food, people can bring their own containers to the store.

Some companies are using a process called loop shopping. First, someone orders an item online. After it arrives, the person leaves the empty container outside. Next, the company picks up the container. Then, the container is used for another order. The loop repeats. As a result, less plastic is used.

Scientists continue to learn about ocean plastic pollution. Some scientists are studying how microplastics affect human health. Others are studying how plastic production contributes to **climate change**.

For many years, plastic pollution in the oceans has gotten worse. But people around the world are working to find new ways to save the oceans from plastic.

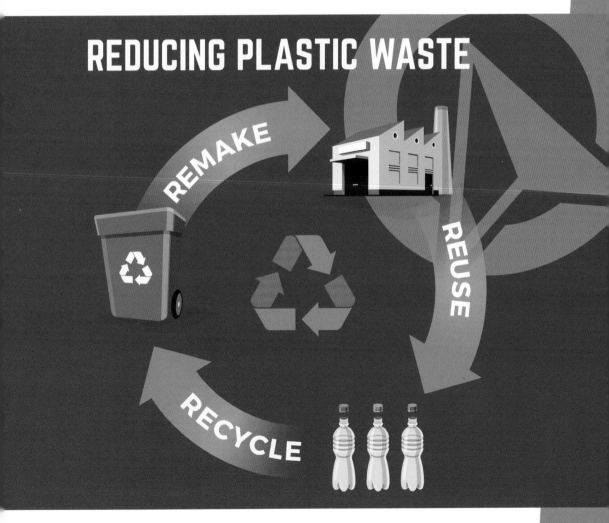

REDUCING PLASTIC WASTE

REMAKE

REUSE

RECYCLE

FOCUS ON
SAVING THE OCEANS FROM PLASTIC

Write your answers on a separate piece of paper.

1. Write a paragraph that describes the key ideas from Chapter 3.

2. Do you think current efforts to save the oceans are enough? Why or why not?

3. What did Ireland do to help reduce plastic pollution?
   - **A.** tax the use of plastic bags
   - **B.** recruit people to clean up ocean garbage patches
   - **C.** make single-use plastics cheaper

4. What might happen if more people refused single-use plastic items?
   - **A.** Climate change might get worse.
   - **B.** More ocean life might die.
   - **C.** Less plastic might enter the oceans.

Answer key on page 32.

GLOSSARY

algae
Tiny, plantlike organisms that live in water and produce oxygen.

biome
A large natural environment with particular communities of plants and animals.

climate change
A human-caused global crisis involving long-term changes in Earth's temperature and weather patterns.

endangered
In danger of dying out.

mammals
Animals that give birth to live babies, have fur or hair, and produce milk.

plankton
Tiny creatures that often drift in big swarms in the ocean.

polymers
Chemical compounds made up of repeating chains of molecules.

species
A group of animals or plants that share the same body shape and can breed with one another.

TO LEARN MORE

BOOKS

Braun, Eric. *Taking Action to Help the Environment.*
  Minneapolis: Lerner Publications, 2017.
LaPierre, Yvette. *Providing Waste Solutions for a City.*
  Minneapolis: Abdo Publishing, 2019.
Smith-Llera, Danielle. *Trash Vortex: How Plastic Pollution
  Is Choking the World's Oceans.* North Mankato, MN:
  Capstone Press, 2018.

NOTE TO EDUCATORS

Visit **www.focusreaders.com** to find lesson plans,
activities, links, and other resources related to this title.

INDEX